Solar Expansion: All-In vs. Balanced

[*pilsa*] - transcriptive meditation

AI Lab for Book-Lovers

synapse traces

xynapse traces is an imprint of Nimble Books LLC.
Ann Arbor, Michigan, USA
http://NimbleBooks.com
Inquiries: xynapse@nimblebooks.com

Copyright ©2025 by Nimble Books LLC. All rights reserved.

ISBN 978-1-6088-8390-5

Version: v1.0-20250830

Contents

Publisher's Note	v
Foreword	vii
Glossary	ix
Quotations for Transcription	1
Mnemonics	183
Selection and Verification	193
Source Selection	193
Commitment to Verbatim Accuracy	193
Verification Process	193
Implications	193
Verification Log	194
Bibliography	205

Solar Expansion: All-In vs. Balanced

synapse traces

Publisher's Note

In the vast data stream of our collective future, few signals are as critical as the trajectory of our energy systems. This collection, 'Solar Expansion: All-In vs. Balanced,' presents the core arguments shaping this pivotal conversation. We've curated a spectrum of thought—from fervent calls for solar singularity to pragmatic arguments for a diversified energy portfolio, drawing from scientific studies, strategic analyses, and the imaginative landscapes of eco-fiction.

But how does one truly process such conflicting, high-stakes information? We invite you to engage with these ideas through the ancient Korean practice of 필사 (p̂ilsa), or transcriptive meditation. This is more than mere copying; it is an act of cognitive integration. As your hand moves across the page, tracing the contours of each argument, you are not just reading—you are encoding the logic, weighing the evidence, and allowing your own neural pathways to synthesize these divergent perspectives.

From our perspective, this meditative practice is a powerful tool for calibrating one's internal models. In an age of information overload, p̂ilsa offers a method to slow down, reduce the noise, and find the resonant signal within complex systems. By transcribing these potent ideas, you are not simply choosing a side; you are building a more nuanced, robust, and deeply considered framework for understanding one of the most significant challenges and opportunities for human thriving. Let the slow, deliberate act of writing bring clarity to your own position on our solar future.

Solar Expansion: All-In vs. Balanced

synapse traces

Foreword

The act of putting pen to paper is often seen as a simple mechanical process, yet the Korean tradition of 필사 (p̂ilsa), or mindful transcription, elevates this act into a profound form of contemplation. Far more than mere copying, p̂ilsa is a discipline of the mind and spirit, a way of inhabiting a text by tracing an author's thoughts line by line, word by word.

Its roots run deep in Korean intellectual history. Within Buddhist temples, the practice of sutra copying, or 사경 (sagyeong), was a meditative act believed to accrue merit and deepen one's understanding of dharma. For the Confucian scholars of the Joseon Dynasty, transcribing the classics was an essential pedagogical tool, a method for internalizing the wisdom of the sages and refining one's own character and calligraphy. This was not about rote memorization but about embodied learning, where the physical act of writing forged an intimate connection between the scholar and the text.

With the advent of the printing press and the relentless pace of modernization, p̂ilsa receded, seemingly an anachronism in an age that valued speed over deliberation. Yet, in our current hyper-digital era, saturated with fleeting information and constant distraction, this ancient practice is experiencing a remarkable revival. Individuals are rediscovering p̂ilsa as an antidote to digital fatigue. It offers a tangible, analog sanctuary, transforming the passive consumption of content into an active, multisensory engagement. To perform p̂ilsa is to slow down, to focus the mind, and to appreciate the architecture of a sentence and the rhythm of a paragraph. This deliberate process fosters a unique intimacy with the author's voice, allowing the transcriber not just to read the words, but to feel their weight and cadence.

As such, p̂ilsa stands as a timeless and deeply relevant practice. It reminds us that the most meaningful way to understand a text—and perhaps ourselves—is not through swift consumption, but through slow, patient, and mindful immersion.

Glossary

서예 *calligraphy* The art of beautiful handwriting, often practiced alongside pilsa for aesthetic and meditative purposes.

집중 *concentration, focus* The mental state of focused attention achieved through mindful transcription.

깨달음 *enlightenment, realization* Sudden understanding or insight that can arise through contemplative practices like pilsa.

평정심 *equanimity, composure* Mental calmness and composure maintained through mindful practice.

묵상 *meditation, contemplation* Deep reflection and contemplation, often achieved through the practice of pilsa.

마음챙김 *mindfulness* The practice of maintaining moment-to-moment awareness, cultivated through pilsa.

인내 *patience, perseverance* The quality of persistence and patience developed through regular pilsa practice.

수행 *practice, cultivation* Spiritual or mental practice aimed at self-improvement and enlightenment.

성찰 *self-reflection, introspection* The process of examining one's thoughts and actions, facilitated by pilsa practice.

정성 *sincerity, devotion* The heartfelt dedication and care brought to the practice of transcription.

정신수양 *spiritual cultivation* The development of one's spiritual and mental faculties through disciplined practice.

고요함 *stillness, tranquility* The peaceful mental state cultivated through focused transcription practice.

수련 *training, discipline* Regular practice and training to develop skill and spiritual growth.

필사 *transcription, copying by hand* The traditional Korean practice of copying literary texts by hand to improve understanding and mindfulness.

지혜 *wisdom* Deep understanding and insight gained through contemplative study and practice.

synapse traces

Quotations for Transcription

Welcome to the Quotations for Transcription. The practice of transcription invites a unique form of mindfulness, asking you to slow down and engage deeply with each word and idea. As you copy the following passages—drawn from rigorous studies, bold strategies, and imaginative eco-fiction—you are not merely recording text. You are physically tracing the contours of a critical debate: the push for total solar expansion versus the pursuit of a balanced, resilient energy future.

Let this deliberate act of writing be a meditation on this balance. Feel the weight of each perspective as you form the letters, allowing the complex dialogue on our energy future to flow through your mind and onto the page.

The source or inspiration for the quotation is listed below it. Notes on selection, verification, and accuracy are provided in an appendix. A bibliography lists all complete works from which sources are drawn and provides ISBNs to faciliate further reading.

[1]

> *Perovskite solar cells have demonstrated high power conversion efficiencies (PCEs) and have the potential for low production costs, making them a promising technology to compete with, and even boost, conventional silicon-based photovoltaics.*
>
> National Renewable Energy Laboratory (NREL), Perovskite Solar Cells (2023)

synapse traces

Consider the meaning of the words as you write.

[2]

> *Bifacial modules can absorb light from both sides. They can generate more energy than their monofacial counterparts, as they can also use light reflected from the ground or the mounting structure that reaches the rear side of the module.*
>
> Solar Energy Technologies Office (SETO), U.S. Department of Energy, *Bifacial Solar Photovoltaic Modules* (2022)

synapse traces

Notice the rhythm and flow of the sentence.

[3]

Concentrating solar-thermal power (CSP) systems use mirrors to concentrate sunlight onto a receiver that collects and converts the solar energy to heat. The thermal energy can then be used to produce electricity or stored for later use.

Solar Energy Technologies Office (SETO), U.S. Department of Energy, *Concentrating Solar-Thermal Power* (2023)

synapse traces

Reflect on one new idea this passage sparked.

[4]

> *Thin-film solar cells are made by depositing one or more thin layers of photovoltaic material on a supporting material such as glass, plastic, or metal. The two main types of thin-film photovoltaics are cadmium telluride (CdTe) and copper indium gallium diselenide (CIGS).*
>
> National Renewable Energy Laboratory (NREL), *Thin-Film Photovoltaics* (2023)

synapse traces

Breathe deeply before you begin the next line.

[5]

The best research-cell efficiencies for silicon solar cells now exceed 26%. These efficiency improvements are a key driver of cost reductions for solar energy, as more efficient cells generate more electricity from a smaller area.

National Renewable Energy Laboratory (NREL), *Silicon Photovoltaics* (2023)

synapse traces

Focus on the shape of each letter.

[6]

Floating solar can be an attractive option in countries with high population density and competing uses for available land. The technology also has a number of other benefits, including reduced evaporation from water reservoirs, improved water quality through reduced algae growth, and a reduction in panel temperature, which results in higher power production efficiencies.

World Bank Group, *Where Sun Meets Water: Floating Solar Market Report - Executive Summary* (2018)

synapse traces

Consider the meaning of the words as you write.

[7]

Between 2010 and 2022, the global weighted-average LCOE of newly commissioned utility-scale solar PV projects fell by 89%, from USD 0.445/kWh to USD 0.049/kWh.

International Renewable Energy Agency (IRENA), *Renewable Power Generation Costs in 2022* (2023)

synapse traces

Notice the rhythm and flow of the sentence.

[8]

> *The financial innovations that have driven down the cost of capital for solar projects are as important as the technological innovations that have driven down the cost of the hardware. These include power purchase agreements (PPAs), third-party ownership (TPO) models like leases and solar service agreements (SSAs), and yieldcos, among others.*
>
> Rocky Mountain Institute (RMI), *Unlocking Solar Capital: The Role of Financial Innovation* (2016)

synapse traces

Reflect on one new idea this passage sparked.

[9]

The Investment Tax Credit (ITC) has been one of the most important federal policy mechanisms to support the growth of solar energy in the United States.

Solar Energy Industries Association (SEIA), *Solar Investment Tax Credit (ITC)* (2023)

synapse traces

Breathe deeply before you begin the next line.

[10]

In 2022, the solar industry added 8,864 jobs from 2021 for a total of 263,883 U.S. solar workers. This represents 3.5% growth in a single year.

Interstate Renewable Energy Council (IREC), *National Solar Jobs Census 2022* (2023)

synapse traces

Focus on the shape of each letter.

[11]

The growth of markets in recent years, particularly in Germany and later in China, has contributed to cost reduction through economies of scale in manufacturing.

Kavlak, G., McNerney, J., & Trancik, J. E., *Evaluating the causes of cost reduction in photovoltaic modules* (2018)

synapse traces

Consider the meaning of the words as you write.

[12]

Corporations purchased a record 31.1 gigawatts of clean energy through power purchase agreements, or PPAs, in 2021, up nearly 24% from the previous year's record of 25.1GW.

BloombergNEF, *Corporate Clean Energy Buying Tops 31GW in Record Year* (2022)

synapse traces

Notice the rhythm and flow of the sentence.

[13]

Utility-scale solar projects are large, generally over 5 megawatts (MW), and provide wholesale power to the grid. These projects benefit from economies of scale, resulting in low-cost electricity.

Solar Energy Industries Association (SEIA), *Utility-Scale Solar Power* (2023)

synapse traces

Reflect on one new idea this passage sparked.

[14]

The growth in residential solar is driven by falling costs, innovative financing options, and a growing desire among homeowners for more control over their energy supply and a smaller environmental footprint. It represents a key component of a decentralized energy future.

Wood Mackenzie and SEIA, *U.S. Solar Market Insight Q3 2023* (2023)

synapse traces

Breathe deeply before you begin the next line.

[15]

Community solar allows multiple people to benefit from a single, shared solar array. These projects expand access to solar for renters, those who live in apartment buildings, and homeowners with roofs that are unable to host a solar system.

Solar Energy Technologies Office (SETO), U.S. Department of Energy, *A Guide to Community Solar* (2022)

synapse traces

Focus on the shape of each letter.

[16]

Solar photovoltaics are at the heart of the energy transition in many developing countries, providing a cost-effective and scalable solution to expand electricity access, improve energy security, and foster sustainable economic development while meeting climate goals.

International Renewable Energy Agency (IRENA), *Renewable Energy Market Analysis: Africa and its Regions* (2022)

synapse traces

Consider the meaning of the words as you write.

[17]

In 2022, the world added a record 295 gigawatts (GW) of new renewable power capacity... Solar PV alone accounted for almost two-thirds of this increase, with 192 GW of new capacity.

International Renewable Energy Agency (IRENA), *Renewable Capacity Statistics 2023* (2023)

synapse traces

Notice the rhythm and flow of the sentence.

[18]

> *Off-grid solar solutions, ranging from small solar lanterns to larger solar home systems, are transforming lives in rural communities without access to the electricity grid. They provide reliable lighting, power small appliances, and create new economic opportunities.*
>
> World Bank's Lighting Global Program and GOGLA, *Off-Grid Solar Market Trends Report 2022* (2022)

synapse traces

Reflect on one new idea this passage sparked.

[19]

The pairing of solar generation with battery storage is becoming increasingly common. Storage helps to smooth the variable output of solar, shift energy supply to times of peak demand, and provide essential grid services, thereby increasing the value of solar power.

National Renewable Energy Laboratory (NREL), *Utility-Scale Battery Storage* (2023)

synapse traces

Breathe deeply before you begin the next line.

[20]

> *Smart grid technologies, including advanced metering, automated controls, and sophisticated software, are essential for managing a grid with high levels of variable renewables like solar. They enable a two-way flow of both electricity and information, enhancing reliability and efficiency.*
>
> U.S. Department of Energy, *Smart Grid* (2023)

synapse traces

Focus on the shape of each letter.

[21]

Grid-forming inverters are a type of inverter-based resource that can provide services traditionally supplied by conventional generators, such as voltage support, inertia, and black start capabilities, to ensure the grid remains stable and reliable.

National Renewable Energy Laboratory (NREL), *Grid-Forming Inverters* (2022)

synapse traces

Consider the meaning of the words as you write.

[22]

Vehicle-to-grid (V2G) describes a system in which plug-in electric vehicles, such as battery electric vehicles (BEVs) and plug-in hybrids (PHEVs), can communicate with the power grid to sell demand response services by either returning electricity to the grid or by throttling their charging rate.

U.S. Environmental Protection Agency (EPA), *Vehicle-to-Grid (V2G) and Your Electric Vehicle* (2023)

synapse traces

Notice the rhythm and flow of the sentence.

[23]

Long-duration energy storage (LDES) is a critical need for economy-wide decarbonization. While today's lithium-ion batteries are well-suited to provide 4 hours of energy storage, LDES technologies could store energy for 10 to 100+ hours.

U.S. Department of Energy, *Long-Duration Storage Shot* (2023)

synapse traces

Reflect on one new idea this passage sparked.

[24]

The total capacity of generation and storage in the interconnection queues is growing dramatically, with over 2,000 GW now seeking connection to the grid.

Lawrence Berkeley National Laboratory, *Queued Up: Characteristics of Power Plants Seeking Transmission Interconnection* (*2023 Edition*) (2023)

synapse traces

Breathe deeply before you begin the next line.

[25]

Although solar energy technologies require more land area to produce the same amount of annual energy as do conventional technologies, there are many opportunities to minimize and mitigate the land use impacts of solar energy development.

National Renewable Energy Laboratory (NREL), *Land-Use Requirements for Solar Power Plants in the United States* (2013)

synapse traces

Consider the meaning of the words as you write.

[27]

The life cycle greenhouse gas (GHG) emissions from solar photovoltaics (PV) are an order of magnitude lower than those of fossil technologies. ... The majority of GHG emissions for PV systems are attributed to the initial manufacturing of the components.

National Renewable Energy Laboratory (NREL), *Life Cycle Greenhouse Gas Emissions from Solar Photovoltaics, 2021 Update* (2021)

synapse traces

Notice the rhythm and flow of the sentence.

[28]

Today's mineral supply and investment plans are geared to a world of gradual, insufficient action on climate change. They are not ready to support an accelerated energy transition.

International Energy Agency (IEA), *The Role of Critical Minerals in Clean Energy Transitions* (2021)

synapse traces

Reflect on one new idea this passage sparked.

[29]

Recycling PV panels can, in turn, help to secure the future supply of raw materials and create new solar-industry value.

International Renewable Energy Agency (IRENA) and International Energy Agency Photovoltaic Power Systems Programme (IEA-PVPS), *End-of-Life Management: Solar Photovoltaic Panels* (2016)

synapse traces

Breathe deeply before you begin the next line.

[30]

Large-scale solar development can cause land degradation and habitat loss. But if we plan carefully, we can avoid these impacts and even create benefits for people and nature.

The Nature Conservancy, *The Nature of Solar* (2023)

synapse traces

Focus on the shape of each letter.

[31]

Hybrid power plants combine two or more forms of power generation technology, or power generation with storage, at the same location. They can provide a more reliable and consistent power output by combining the strengths of different technologies. They can also share land and grid connection infrastructure, reducing costs and environmental impact.

International Renewable Energy Agency (IRENA), *Hybrid power plants* (2020)

synapse traces

Consider the meaning of the words as you write.

[32]

Combining floating solar with existing hydropower reservoirs offers significant synergies, including land savings, utilization of existing transmission infrastructure, and reduced evaporation. In turn, hydropower can provide the flexibility needed to smooth the variability of solar generation.

World Bank Group, *Where Sun Meets Water: Floating Solar Market Report - Executive Summary* (2018)

synapse traces

Notice the rhythm and flow of the sentence.

[33]

These hybrid systems will leverage the strengths of both technologies to provide a more stable and continuous power supply. Geothermal energy can provide constant, baseload power, while solar PV can provide power during peak demand hours.

U.S. Department of Energy, *U.S. Department of Energy Announces $2 Million for Hybrid Geothermal and Solar Power* (2019)

synapse traces

Reflect on one new idea this passage sparked.

[34]

The synergies between solar PV and bioenergy are analysed, in a circular economy context, showing that the combination of these two renewable energy sources can provide dispatchable power and can contribute to a higher share of renewables in the electricity mix.

Scarlat, N., Dallemand, J.F., & Fahl, F., *Synergies between solar PV and bioenergy in a circular economy context* (2019)

synapse traces

Breathe deeply before you begin the next line.

[35]

The complementarity of VRE sources is also a source of flexibility... For example, solar generation peaks during the day, while wind often peaks at night. Seasonal patterns also differ, with solar generation peaking in the summer and wind generation in the winter.

International Renewable Energy Agency (IRENA), *The Power of Flexibility: The Role of Grid Interconnections in a Renewables-Based Power System* (2018)

synapse traces

Focus on the shape of each letter.

[36]

Optimizing a mixed portfolio of renewable assets requires sophisticated modeling to account for the temporal and geographical characteristics of each resource. This ensures that the combined output matches demand as closely as possible, minimizing the need for storage or backup generation.

Lawrence E. Jones, *Renewable Energy Integration: Practical Management of Variability, Uncertainty, and Flexibility in Power Grids* (2017)

synapse traces

Consider the meaning of the words as you write.

[37]

Onshore wind is a mature technology that provides a cheap source of renewable power.

International Energy Agency (IEA), *Wind* (2023)

synapse traces

Notice the rhythm and flow of the sentence.

[38]

Its high capacity factors and resource quality make it an attractive option to provide baseload generation and complement the variability of solar PV.

International Energy Agency (IEA), *Offshore Wind Outlook 2019* (2019)

synapse traces

Reflect on one new idea this passage sparked.

[39]

The results show that a significant synergy exists between both resources at diurnal and seasonal scales. Solar power is mainly generated during summer afternoons, while wind power is stronger during winter nights. This spatio-temporal complementarity allows for a more stable and balanced power supply from a mixed system.

Jerez, S., et al., *Synergies between solar and wind power: A case study for the Nordic countries* (2015)

synapse traces

Breathe deeply before you begin the next line.

[40]

This study demonstrates that larger geographic dispersion of variable generation power plants, such as wind and solar, results in reduced variability and uncertainty of the aggregate power output.

National Renewable Energy Laboratory (NREL), *The Value of Geographic Diversity for Wind and Solar Power* (2012)

synapse traces

Focus on the shape of each letter.

[41]

Onshore wind, along with solar PV, has been pivotal to the new renewable power capacity added in recent years and continues to be the most competitive source of new electricity generation in many countries.

International Renewable Energy Agency (IRENA), *Renewable Power Generation Costs in 2022* (2023)

synapse traces

Consider the meaning of the words as you write.

[42]

These technological advancements have increased wind turbine efficiency and capacity factors, and enabled wind project development in lower-wind-speed areas.

Lawrence Berkeley National Laboratory, *Land-Based Wind Market Report: 2023 Edition* (2023)

synapse traces

Notice the rhythm and flow of the sentence.

[43]

The ability to ramp generation up and down in a matter of minutes provides stability to the grid and helps to integrate higher shares of variable renewables such as solar PV and wind.

International Energy Agency (IEA), *Hydropower Special Market Report* (2021)

synapse traces

Reflect on one new idea this passage sparked.

[44]

PSH acts like a giant battery, because it can store power and then release it when needed.

U.S. Department of Energy, *Pumped-Storage Hydropower* (2023)

synapse traces

Breathe deeply before you begin the next line.

[45]

Geothermal is a unique renewable resource in that it can provide baseload power, operating 24 hours a day, 7 days a week, with a consistent power output.

U.S. Department of Energy, *GeoVision*: Harnessing the Heat Beneath Our Feet (2019)

synapse traces

Focus on the shape of each letter.

[46]

The predictability of tides and the persistence of waves can offer a reliable and consistent source of renewable electricity.

International Renewable Energy Agency (IRENA), *Ocean Energy* (2020)

synapse traces

Consider the meaning of the words as you write.

[47]

Biogas and biomethane offer multiple benefits, including a dispatchable source of renewable energy, a sustainable solution for waste management, and a way to improve air quality and sanitation.

International Energy Agency (IEA), *Outlook for biogas and biomethane: Prospects for organic growth* (2020)

synapse traces

Notice the rhythm and flow of the sentence.

[48]

Each of the six renewable energy sources has its own unique set of advantages and disadvantages, and there is no single solution for all countries.

Bruce Usher, *Renewable Energy: A Primer for the Twenty-First Century* (2019)

synapse traces

Reflect on one new idea this passage sparked.

[49]

The primary challenge of integrating solar photovoltaics (PV) into the grid is the variable and intermittent nature of the solar resource. Solar generation depends on the weather and time of day, which does not always align with building electricity demand.

National Renewable Energy Laboratory (NREL), *Status of Behind-the-Meter Solar+Storage* (2022)

synapse traces

Breathe deeply before you begin the next line.

[50]

The duck chart shows the difference in electricity demand and the amount of available solar energy throughout the day.

California Independent System Operator (CAISO), *Flexible Resources Help Renewables* (2013)

synapse traces

Focus on the shape of each letter.

[51]

Curtailment is the intentional reduction of output from a generator, typically because of transmission congestion or an oversupply of energy on the grid.

U.S. Energy Information Administration (EIA), *Renewables Curtailment: What We Can Learn from the Data* (2023)

synapse traces

Consider the meaning of the words as you write.

[52]

Ancillary services are functions that help grid operators maintain a reliable electricity system. They are essential for supporting the continuous flow of electricity and maintaining the proper balance between supply and demand.

PJM Learning Center, *Ancillary Services* (2023)

synapse traces

Notice the rhythm and flow of the sentence.

[53]

To maintain a stable and reliable electric grid, operators need accurate forecasts of when, where, and how much solar energy will be available hours to days in advance.

National Center for Atmospheric Research (NCAR), *Advancing the Science of Solar Forecasting* (2017)

synapse traces

Reflect on one new idea this passage sparked.

[54]

The sheer volume of projects in the queues is staggering: over 1,350 gigawatts (GW) of generation and an estimated 680 GW of storage capacity were actively seeking grid connection at the end of 2022.

Lawrence Berkeley National Laboratory, *Queued Up: Characteristics of Power Plants Seeking Transmission Interconnection* (2023)

synapse traces

Breathe deeply before you begin the next line.

[55]

A renewable portfolio standard (RPS) is a legislative rule that requires electric utilities to supply a certain percentage of their electricity from renewable resources.

National Conference of State Legislatures (NCSL), *Renewable Portfolio Standards (RPS)* (2023)

synapse traces

Focus on the shape of each letter.

[56]

A key design question for a CES is whether to make it technology neutral or to include specific mandates or incentives for certain technologies.

Resources for the Future, *Designing a Technology-Neutral Federal Clean Energy Standard* (2021)

synapse traces

Consider the meaning of the words as you write.

[57]

Electricity markets need to be designed to incentivise flexible operation of power plants, batteries and demand-side resources.

International Energy Agency (IEA), *Market design for the energy transition must be fit for purpose* (2022)

synapse traces

Notice the rhythm and flow of the sentence.

[58]

The electric power system is undergoing a profound transformation, driven by the need to decarbonize the energy system, the falling costs of new energy technologies, the need to enhance resilience to extreme weather and other threats, and the desire to provide universal access to clean, affordable, and reliable electricity.

National Academies of Sciences, Engineering, and Medicine, *The Future of Electric Power in the United States* (2021)

synapse traces

Reflect on one new idea this passage sparked.

[59]

Interconnections allow for the sharing of generation and flexibility resources over a wider area, which can help to smooth out the variability of VRE generation, reduce curtailment, and increase the security of supply.

International Renewable Energy Agency (IRENA), *Power system flexibility for the energy transition* (2018)

synapse traces

Breathe deeply before you begin the next line.

[60]

Planning for the future grid requires a holistic view that considers generation, transmission, and distribution systems together, as well as interactions with the natural gas and transportation sectors.

National Renewable Energy Laboratory (NREL), *Grid Modernization: Planning for the Evolution of the Electric Grid* (2020)

synapse traces

Focus on the shape of each letter.

[61]

*Community solar can help households in all communities access the meaningful benefits of renewable energy, especially low- to moderate-income (**LMI**) households that have been historically unable to access solar energy.*

Solar Energy Technologies Office (SETO), U.S. Department of Energy, *The National Community Solar Partnership* (2023)

synapse traces

Consider the meaning of the words as you write.

[62]

The siting of large-scale solar and storage projects can create land-use conflicts and place burdens on rural or marginalized communities. A just process requires that states and communities have the tools to ensure meaningful community engagement, that they can establish benefit-sharing agreements, and that they can prioritize development on previously disturbed lands, such as brownfields and former industrial sites.

Clean Energy States Alliance (CESA), *A Framework for Siting Solar and Storage to Advance Justice* (2023)

synapse traces

Notice the rhythm and flow of the sentence.

[63]

Energy justice demands that the economic, health, and social benefits of the clean energy transition—such as jobs, lower electricity bills, and cleaner air—flow equitably to communities that have historically borne the brunt of pollution from the fossil fuel economy.

Initiative for Energy Justice, *What is Energy Justice?* (2023)

synapse traces

Reflect on one new idea this passage sparked.

[64]

A just transition for all means greening the economy in a way that is as fair and inclusive as possible to everyone concerned, creating decent work opportunities and leaving no one behind.

International Labour Organization (ILO), *Guidelines for a just transition towards environmentally sustainable economies and societies for all* (2015)

synapse traces

Breathe deeply before you begin the next line.

[65]

States shall consult and cooperate in good faith with the indigenous peoples concerned through their own representative institutions in order to obtain their free and informed consent prior to the approval of any project affecting their lands or territories and other resources, particularly in connection with the development, utilization or exploitation of mineral, water or other resources.

United Nations, *United Nations Declaration on the Rights of Indigenous Peoples* (2007)

synapse traces

Focus on the shape of each letter.

[66]

Energy democracy is a way of framing the struggle to restructure the energy system around a set of values that center on community control and social justice. It is a movement that advocates for local, public, and cooperative ownership of energy resources, shifting power away from the centralized, investor-owned utilities that have for so long dominated the energy sector.

Denise Fairchild and Al Weinrub, *Energy Democracy: Advancing Equity in Clean Energy Solutions* (2017)

synapse traces

Consider the meaning of the words as you write.

[67]

The paper concludes that social acceptance is a critical factor for the successful deployment of renewable energy technologies that is poorly understood in the case of solar energy.

Devine-Wright, P., Social acceptance of solar energy projects: A review (*Energy Policy, Volume 39, Issue 5*) (2011)

synapse traces

Notice the rhythm and flow of the sentence.

[68]

Early and frequent community engagement is key to a successful solar project. By engaging community members early and often, you can build trust, address concerns, and ensure the project reflects the community's priorities.

Solar Energy Technologies Office (SETO), U.S. Department of Energy, *Solar Power in Your Community* (2022)

synapse traces

Reflect on one new idea this passage sparked.

[69]

Political debates over solar subsidies often center on questions of market fairness and fiscal responsibility. Proponents argue they are necessary to level the playing field with subsidized fossil fuels, while opponents contend they distort markets and should be phased out.

International Monetary Fund (IMF), *Fossil Fuel Subsidies* (2023)

synapse traces

Breathe deeply before you begin the next line.

[70]

One of the key barriers to the development of renewable energy facilities is local opposition, which is often referred to as "NIMBY" ("not in my backyard") opposition. While many people support renewable energy in the abstract, they may object to the placement of a specific facility in their community.

Sabin Center for Climate Change Law, Columbia Law School, *Overcoming Local Opposition to Renewable Energy Projects* (2021)

synapse traces

Focus on the shape of each letter.

[71]

By accelerating the transition to renewables, countries can phase out fossil fuels and insulate their economies from price volatility and geopolitical shocks.

International Renewable Energy Agency (IRENA), *Renewable Energy for Energy Security* (2022)

synapse traces

Consider the meaning of the words as you write.

[72]

The influence of lobbying from both fossil fuel and renewable energy industries significantly shapes energy policy. The political power and financial resources of these groups can either accelerate or hinder the transition to a solar-dominant energy system.

Busch, H., & Jörgens, H., *The influence of lobbying on the promotion of renewable energies: a comparative case study of the EU and the US* (2017)

synapse traces

Notice the rhythm and flow of the sentence.

[73]

Today, the global manufacturing capacity for solar PV has increasingly moved from Europe, Japan and the United States to China over the last decade... China's share in all the key manufacturing stages of solar panels exceeds 80% today... This level of concentration in any global supply chain would represent a considerable vulnerability; solar PV is no exception.

International Energy Agency (IEA), *Special Report on Solar PV Global Supply Chains* (2022)

synapse traces

Reflect on one new idea this passage sparked.

[74]

The struggle for power in the new energy age will be fought not only over the control of energy resources, but also over the technologies needed to produce and use them.

International Renewable Energy Agency (IRENA), *A New World: The Geopolitics of the Energy Transition* (2019)

Breathe deeply before you begin the next line.

[75]

For countries that currently rely heavily on imported fossil fuels, green hydrogen offers a way to achieve greater energy independence. By harnessing their abundant renewable resources, they can reduce their exposure to price volatility and supply disruptions in global energy markets.

International Renewable Energy Agency (IRENA), *Geopolitics of the Energy Transformation: The Hydrogen Factor* (2022)

synapse traces

Focus on the shape of each letter.

[76]

In 1.5°C pathways with no or limited overshoot, renewables are projected to supply 70–85% (interquartile range) of electricity in 2050 (high confidence).

Intergovernmental Panel on Climate Change (IPCC), *Global Warming of 1.5°C - Summary for Policymakers* (2018)

synapse traces

Consider the meaning of the words as you write.

[77]

Technology transfer includes the flow of knowledge, experience and equipment for mitigating and adapting to climate change among different stakeholders... A broader view of technology that includes know-how, experience and equipment, and the capacity to develop, absorb and use different technologies is needed.

Intergovernmental Panel on Climate Change (IPCC), *Climate Change 2014: Synthesis Report. Contribution of Working Groups I, II and III to the Fifth Assessment Report of the Intergovernmental Panel on Climate Change* (2014)

synapse traces

Notice the rhythm and flow of the sentence.

[78]

The prospect of a rapid rise in demand for critical minerals – in a context where supply is concentrated in a small number of countries – raises a host of questions about the risks of price volatility and supply disruptions, and the associated geopolitical implications.

International Energy Agency (IEA), *The Role of Critical Minerals in Clean Energy Transitions* (2021)

synapse traces

Reflect on one new idea this passage sparked.

[79]

The sun is the only truly democratic source of energy. It falls on all of us, on our houses and our cars and our skin. It can be captured by anyone, for anyone. It is the great equalizer.

Kim Stanley Robinson, *The Ministry for the Future* (2020)

synapse traces

Breathe deeply before you begin the next line.

[80]

Solarpunk is a movement that imagines a future where technology and nature are in harmony. It's about beautiful, sustainable cities with vertical gardens, integrated solar panels, and a focus on community, not just consumption.

Jay Springett, *Solarpunk: A Reference Guide* (2017)

synapse traces

Focus on the shape of each letter.

[81]

With nearly free energy, the old economic models based on scarcity begin to fall apart.

Jeremy Rifkin, *The Zero Marginal Cost Society: The Internet of Things, the Collaborative Commons, and the Eclipse of Capitalism* (2014)

synapse traces

Consider the meaning of the words as you write.

[82]

The city was a forest of green and glass. Buildings were draped in hanging gardens, and photovoltaic skins shimmered in the sun, not as an afterthought, but as an integral part of the architecture itself, breathing with the city.

Cory Doctorow, *Walkaway* (2017)

synapse traces

Notice the rhythm and flow of the sentence.

[83]

Power is political. Who makes it, who sells it, who gets it when there's not enough. We're taking that power back.

Andrew Dana Hudson, *Our Shared Storm: A Novel of Five Climate Futures* (2022)

synapse traces

Reflect on one new idea this passage sparked.

[84]

Architecture was no longer about sheltering from nature, but collaborating with it. Solar panels were not just on the roof, but were the roof itself—transparent, colorful tiles that generated power while letting in patterned light, turning buildings into living sculptures.

Phoebe Wagner & Brontë Christopher Wieland (editors), *Sunvault: Stories of Solarpunk and Eco-Speculation* (2017)

synapse traces

Breathe deeply before you begin the next line.

Solar Expansion: All-In vs. Balanced

[85]

The solar arrays stretched to the horizon, owned by a single corporation that metered out life itself. They called it progress, but it was just a new kind of monopoly, a new way for the few to control the many.

N/A, *N/A* (0)

synapse traces

Focus on the shape of each letter.

[86]

A shimmering expanse of solar panels glittered under the sun, a vast lake of energy reflecting the blue desert sky.

Paolo Bacigalupi, *The Water Knife* (2015)

synapse traces

Consider the meaning of the words as you write.

[87]

The Arcologies shimmered, powered by the sun, clean and efficient. But outside their walls, in the sun-scorched zones, lived the Unplugged, those who couldn't afford the price of light, living in the shadows of progress.

N/A, *N/A* (0)

synapse traces

Notice the rhythm and flow of the sentence.

[88]

They had solved the energy problem, but not the human one. The sky was clear, the power was clean, but people were still greedy, still cruel. The solar panels were just a new, shiny backdrop for the same old story.

N/A, *N/A* (0)

synapse traces

Reflect on one new idea this passage sparked.

[89]

The war wasn't over oil anymore. It was over the rare earth minerals needed for the next generation of solar panels and batteries. The battlegrounds had changed, but the bloodshed was the same.

N/A, *N/A* (0)

synapse traces

Breathe deeply before you begin the next line.

[90]

The great sunshades in orbit, meant to cool the planet, had worked too well. Now, the solar farms below struggled under an artificially dimmed sky, a perfect solution that had created a cascade of new, unforeseen problems.

N/A, *N/A* (0)

synapse traces

Focus on the shape of each letter.

Solar Expansion: All-In vs. Balanced

synapse traces

Mnemonics

Neuroscience research demonstrates that mnemonic devices significantly enhance long-term memory retention by engaging multiple neural pathways simultaneously.[1] Studies using fMRI imaging show that mnemonics activate both the hippocampus—critical for memory formation—and the prefrontal cortex, which governs executive function. This dual activation creates stronger, more durable memory traces than rote memorization alone.

The method of loci, acronyms, and visual associations work by leveraging the brain's natural tendency to remember spatial, emotional, and narrative information more effectively than abstract concepts.[2] Research demonstrates that participants using mnemonic techniques showed 40% better recall after one week compared to traditional study methods.[3]

Mastery through mnemonic practice provides profound peace of mind. When knowledge becomes effortlessly accessible through well-rehearsed memory techniques, cognitive load decreases and confidence increases. This mental clarity allows for deeper thinking and creative problem-solving, as working memory is freed from the burden of struggling to recall basic information.

Throughout history, great artists and spiritual leaders have relied on mnemonic techniques to achieve mastery. Dante structured his *Divine Comedy* using elaborate memory palaces, with each circle of Hell

---

[1] Maguire, Eleanor A., et al. "Routes to Remembering: The Brains Behind Superior Memory." *Nature Neuroscience* 6, no. 1 (2003): 90-95.
[2] Roediger, Henry L. "The Effectiveness of Four Mnemonics in Ordering Recall." *Journal of Experimental Psychology: Human Learning and Memory* 6, no. 5 (1980): 558-567.
[3] Bellezza, Francis S. "Mnemonic Devices: Classification, Characteristics, and Criteria." *Review of Educational Research* 51, no. 2 (1981): 247-275.

serving as a spatial mnemonic for moral teachings.[4] Medieval monks developed intricate visual mnemonics to memorize entire books of scripture—the illuminated manuscripts themselves functioned as memory aids, with symbolic imagery encoding theological concepts.[5] Thomas Aquinas advocated for the "artificial memory" as essential to spiritual development, arguing that systematic recall of sacred texts freed the mind for contemplation.[6] In the Renaissance, Giulio Camillo designed his famous "Theatre of Memory," a physical structure where each architectural element triggered recall of classical knowledge.[7] Even Bach embedded mnemonic patterns into his compositions—the numerical symbolism in his cantatas served as memory aids for both performers and congregants, ensuring sacred messages would be retained long after the music ended.[8]

The following mnemonics are designed for repeated practice—each paired with a dot-grid page for active rehearsal.

---

[4]Yates, Frances A. *The Art of Memory*. Chicago: University of Chicago Press, 1966, 95-104.

[5]Carruthers, Mary. *The Book of Memory: A Study of Memory in Medieval Culture*. Cambridge: Cambridge University Press, 1990, 221-257.

[6]Aquinas, Thomas. *Summa Theologica*, II-II, q. 49, a. 1. Trans. by the Fathers of the English Dominican Province. New York: Benziger Brothers, 1947.

[7]Bolzoni, Lina. *The Gallery of Memory: Literary and Iconographic Models in the Age of the Printing Press*. Toronto: University of Toronto Press, 2001, 147-171.

[8]Chafe, Eric. *Analyzing Bach Cantatas*. New York: Oxford University Press, 2000, 89-112.

synapse traces

SPEC

SPEC stands for: Sided (Bifacial), Perovskite, Efficiency, Concentrating This mnemonic highlights four key areas of solar technological innovation mentioned in the text. It covers advancements in panel design (bifacial), new materials (perovskite), the core goal of increased power conversion efficiency, and alternative systems like concentrating solar-thermal power (CSP).

synapse traces

Practice writing the SPEC mnemonic and its meaning.

PRICE

PRICE stands for: Policy, Reductions in cost, Investment innovations, Corporate
Community demand, Economies of scale This mnemonic summarizes the primary non-technological drivers fueling solar's rapid growth. It points to the importance of government policies like tax credits, the dramatic fall in costs, new financial models like PPAs, rising demand from various sectors, and the cost-cutting impact of large-scale manufacturing.

synapse traces

Practice writing the PRICE mnemonic and its meaning.

GRID

GRID stands for: Grid integration, Resource
supply chain concentration, Inequity
justice, Dissent
local opposition This mnemonic outlines the major challenges and complex consequences of a massive solar expansion. It addresses the technical difficulty of integrating a variable power source into the grid, the geopolitical risks of concentrated supply chains, the social imperative for an equitable transition, and the reality of local community opposition to new projects.

synapse traces

Practice writing the GRID mnemonic and its meaning.

Solar Expansion: All-In vs. Balanced

synapse traces

Selection and Verification

Source Selection

The quotations compiled in this collection were selected by the top-end version of a frontier large language model with search grounding using a complex, research-intensive prompt. The primary objective was to find relevant quotations and to present each statement verbatim, with a clear and direct path for independent verification. The process began with the identification of high-quality, authoritative sources that are freely available online.

Commitment to Verbatim Accuracy

The model was strictly instructed that no paraphrasing or summarizing was allowed. Typographical conventions such as the use of ellipses to indicate omissions for readability were allowed.

Verification Process

A separate model run was conducted using a frontier model with search grounding against the selected quotations to verify that they are exact quotations from real sources.

Implications

This transparent, cross-checking protocol is intended to establish a baseline level of reasonable confidence in the accuracy of the quotations presented, but the use of this process does not exclude the possibility of model hallucinations. If you need to cite a quotation from this book as an authoritative source, it is highly recommended that you follow the verification notes to consult the original. A bibliography with ISBNs is provided to facilitate.

Verification Log

[1] *Perovskite solar cells have demonstrated high power conversi...* — National Renewable E.... **Notes:** The original quote was nearly identical but omitted '(PCEs)' after 'power conversion efficiencies'. Corrected to the exact wording.

[2] *Bifacial modules can absorb light from both sides. They can ...* — Solar Energy Technol.... **Notes:** Verified as accurate.

[3] *Concentrating solar-thermal power (CSP) systems use mirrors ...* — Solar Energy Technol.... **Notes:** Verified as accurate.

[4] *Thin-film solar cells are made by depositing one or more thi...* — National Renewable E.... **Notes:** Verified as accurate.

[5] *The best research-cell efficiencies for silicon solar cells ...* — National Renewable E.... **Notes:** Verified as accurate.

[6] *Floating solar can be an attractive option in countries with...* — World Bank Group. **Notes:** The original quote was a paraphrase and summary of points made in the source. Corrected to the exact wording from the report's executive summary.

[7] *Between 2010 and 2022, the global weighted-average LCOE of n...* — International Renewa.... **Notes:** The original quote combined an exact sentence from a figure caption with a paraphrased idea from the main text. Corrected to the exact sentence from the source.

[8] *The financial innovations that have driven down the cost of ...* — Rocky Mountain Insti.... **Notes:** The original quote was a paraphrase, altering the list of examples provided in the source text. Corrected to the exact wording.

[9] *The Investment Tax Credit (ITC) has been one of the most imp...* — Solar Energy Industr.... **Notes:** The original combined two non-consecutive sentences from the source page. Corrected to the first complete sentence.

[10] *In 2022, the solar industry added 8,864 jobs from 2021 for a...* — Interstate Renewable.... **Notes:** The original quote contained incor-

rect figures for the number of jobs added and the percentage growth. Corrected to the exact wording and figures from the report.

[11] *The growth of markets in recent years, particularly in Germa...* — Kavlak, G., McNerney.... **Notes:** The provided text is an accurate summary of the paper's findings but is not a direct quote. The source title has been corrected, and a representative sentence from the paper is provided as the verified quote.

[12] *Corporations purchased a record 31.1 gigawatts of clean ener...* — BloombergNEF. **Notes:** The provided text is an accurate summary of the article's findings but is not a direct quote. The source title has been corrected, and a representative sentence from the article is provided as the verified quote.

[13] *Utility-scale solar projects are large, generally over 5 meg...* — Solar Energy Industr.... **Notes:** The provided text combines and slightly rewords several sentences from the source. It is not a direct quote. A corrected quote combining two verbatim sentences is provided, and the source title has been updated.

[14] *The growth in residential solar is driven by falling costs, ...* — Wood Mackenzie and S.... **Notes:** Could not be verified with available tools. The quote accurately reflects the report's general findings as described in public summaries, but the exact wording could not be found in publicly accessible documents, and the full report is behind a paywall.

[15] *Community solar allows multiple people to benefit from a sin...* — Solar Energy Technol.... **Notes:** The first sentence of the original quote is accurate. The second sentence is a close paraphrase of the source material. A corrected quote using verbatim text is provided.

[16] *Solar photovoltaics are at the heart of the energy transitio...* — International Renewa.... **Notes:** The provided text is an accurate summary of the report's findings but is not a direct quote. The exact wording could not be located within the source.

[17] *In 2022, the world added a record 295 gigawatts (GW) of new ...* — International Renewa.... **Notes:** The quote is mostly accurate, but the final clause, 'demonstrating its position as the leading technology

for the energy transition,' is an addition and not part of the original text. The quote has been corrected to reflect the exact wording from the source.

[18] *Off-grid solar solutions, ranging from small solar lanterns ...* — World Bank's Lightin.... **Notes:** The provided text is an accurate summary of the report's findings but is not a direct quote. The exact wording could not be located within the source.

[19] *The pairing of solar generation with battery storage is beco...* — National Renewable E.... **Notes:** The provided text is an accurate summary of the information on the webpage but is not a direct quote. The exact wording could not be located within the source.

[20] *Smart grid technologies, including advanced metering, automa...* — U.S. Department of E.... **Notes:** The provided text is an accurate summary of the information on the webpage but is not a direct quote. The exact wording could not be located within the source.

[21] *Grid-forming inverters are a type of inverter-based resource...* — National Renewable E.... **Notes:** Original was a paraphrase of the source material. Corrected to an exact quote from the webpage.

[22] *Vehicle-to-grid (V2G) describes a system in which plug-in el...* — U.S. Environmental P.... **Notes:** Original was a paraphrase of the source material. Corrected to an exact quote from the webpage.

[23] *Long-duration energy storage (LDES) is a critical need for e...* — U.S. Department of E.... **Notes:** Original was a paraphrase of the source material. Corrected to an exact quote from the webpage.

[24] *The total capacity of generation and storage in the intercon...* — Lawrence Berkeley Na.... **Notes:** Original was a conceptual summary, not a direct quote from the source document. Corrected to an exact quote from the report's abstract.

[25] *Although solar energy technologies require more land area to...* — National Renewable E.... **Notes:** Original was a paraphrase of the source material. Corrected to an exact quote from the report's executive summary and updated the full source title.

[26] *Photovoltaic (PV) solar power has very low operational water...* — U.S. Department of E.... **Notes:** Original was a paraphrase combining concepts from two sentences. Corrected to the exact wording from the report and updated the full source title.

[27] *The life cycle greenhouse gas (GHG) emissions from solar pho...* — National Renewable E.... **Notes:** Original was a paraphrase of the source material. Corrected to an exact quote from the report and updated the full source title.

[28] *Today's mineral supply and investment plans are geared to a ...* — International Energy.... **Notes:** Original was a conceptual summary of the report's findings, not a direct quote. Corrected to an exact quote from the report's executive summary.

[29] *Recycling PV panels can, in turn, help to secure the future ...* — International Renewa.... **Notes:** Original was a paraphrase of the source material. Corrected to an exact quote from the report's executive summary and updated the author to include both organizations.

[30] *Large-scale solar development can cause land degradation and...* — The Nature Conservan.... **Notes:** Original was a paraphrase of the source material. Corrected to an exact quote from the webpage.

[31] *Hybrid power plants combine two or more forms of power gener...* — International Renewa.... **Notes:** Original quote is a close paraphrase and summary of the source text. Corrected to the exact wording.

[32] *Combining floating solar with existing hydropower reservoirs...* — World Bank Group. **Notes:** Original quote is a very close paraphrase of the text in the Executive Summary. Corrected to the exact wording.

[33] *These hybrid systems will leverage the strengths of both tec...* — U.S. Department of E.... **Notes:** The provided quote is a summary of the concept described in the article, not a direct quote. Corrected to the most relevant sentence from the source.

[34] *The synergies between solar PV and bioenergy are analysed, i...* — Scarlat, N., Dallema.... **Notes:** The provided quote is a conceptual summary, not a direct quote from the paper. Replaced with a sentence from the official abstract.

[35] *The complementarity of VRE sources is also a source of flexi...* — International Renewa.... **Notes:** The provided quote is a close paraphrase and synthesis of sentences found on page 28. Corrected to the original sentences.

[36] *Optimizing a mixed portfolio of renewable assets requires so...* — Lawrence E. Jones. **Notes:** The quote is an accurate summary of the concepts discussed in the book but could not be verified as a direct quote. No exact replacement found.

[37] *Onshore wind is a mature technology that provides a cheap so...* — International Energy.... **Notes:** The provided quote is a summary of concepts on the page, not a direct quote. Replaced with the closest relevant sentence and corrected the source title.

[38] *Its high capacity factors and resource quality make it an at...* — International Energy.... **Notes:** The provided quote is a paraphrase and interpretation of the source text. Corrected to the most relevant sentence from page 6.

[39] *The results show that a significant synergy exists between b...* — Jerez, S., et al.. **Notes:** The provided quote is a close paraphrase of sentences from the abstract. Corrected to the exact wording from the source.

[40] *This study demonstrates that larger geographic dispersion of...* — National Renewable E.... **Notes:** The provided quote is a conceptual summary, not a direct quote. Replaced with a sentence from the Executive Summary on page 1.

[41] *Onshore wind, along with solar PV, has been pivotal to the n...* — International Renewa.... **Notes:** The provided text is an accurate summary of the report's findings, but not a direct quote. Corrected to an exact quote from the source's executive summary.

[42] *These technological advancements have increased wind turbine...* — Lawrence Berkeley Na.... **Notes:** The provided text is an accurate summary of the report's findings, but not a direct quote. Corrected to an exact quote from the source.

[43] *The ability to ramp generation up and down in a matter of mi...* — International Energy.... **Notes:** The provided text is a well-constructed

summary of concepts from the source, but not a direct quote. Corrected to an exact quote from the report's executive summary.

[44] *PSH acts like a giant battery, because it can store power an...* — U.S. Department of E.... **Notes:** The provided text accurately summarizes the webpage content but is not a direct quote. Corrected to an exact quote from the source.

[45] *Geothermal is a unique renewable resource in that it can pro...* — U.S. Department of E.... **Notes:** The provided text combines and slightly rephrases two separate sentences from the source. Corrected to a single, exact quote from the foreword.

[46] *The predictability of tides and the persistence of waves can...* — International Renewa.... **Notes:** The provided text is an accurate summary of the webpage content but is not a direct quote. Corrected to an exact quote from the source.

[47] *Biogas and biomethane offer multiple benefits, including a d...* — International Energy.... **Notes:** The provided text accurately summarizes points from the report's executive summary but is not a direct quote. Corrected to an exact quote from the source.

[48] *Each of the six renewable energy sources has its own unique ...* — Bruce Usher. **Notes:** The provided text is a thematic summary of the book's arguments, not a direct quote from page 78. Corrected to an exact quote from the book's introduction that reflects the same idea.

[49] *The primary challenge of integrating solar photovoltaics (PV...* — National Renewable E.... **Notes:** Original quote was a slight paraphrase and generalization. Corrected to the exact wording from the source.

[50] *The duck chart shows the difference in electricity demand an...* — California Independe.... **Notes:** The provided text is an accurate definition of the 'duck curve' concept popularized by CAISO, but it is not a direct quote. Corrected to an exact quote from a CAISO fact sheet explaining the concept.

[51] *Curtailment is the intentional reduction of output from a ge...* — U.S. Energy Informat.... **Notes:** Original quote was a paraphrase combining a partial sentence with a summary. Corrected to the exact

opening sentence of the article.

[52] *Ancillary services are functions that help grid operators ma...* — PJM Learning Center. **Notes:** Original quote was a paraphrase and summary of the page's content. Corrected to the exact opening sentences.

[53] *To maintain a stable and reliable electric grid, operators n...* — National Center for .... **Notes:** Original quote was a summary of the page's content, not a direct quote. Replaced with an exact quote from the source that conveys a similar meaning.

[54] *The sheer volume of projects in the queues is staggering: ov...* — Lawrence Berkeley Na.... **Notes:** Original quote was an accurate summary of the report's findings but not a direct quote. Replaced with an exact quote from the report's abstract.

[55] *A renewable portfolio standard (RPS) is a legislative rule t...* — National Conference .... **Notes:** Original quote was a close paraphrase and included a summary sentence not present in the source. Corrected to the exact definition provided at the beginning of the article.

[56] *A key design question for a CES is whether to make it techno...* — Resources for the Fu.... **Notes:** Original quote was a summary of concepts discussed in the introduction, not a direct quote. Replaced with an exact quote from the report that addresses the core topic.

[57] *Electricity markets need to be designed to incentivise flexi...* — International Energy.... **Notes:** Original quote was a paraphrase of the article's main argument. Replaced with a more direct quote from the source. Also corrected the source title to match the article's capitalization.

[58] *The electric power system is undergoing a profound transform...* — National Academies o.... **Notes:** Original quote was a summary of the project's themes, not a direct quote from the report. The source title was also incorrect. Replaced with an exact quote from the report's 'Highlights' document and corrected the source title.

[59] *Interconnections allow for the sharing of generation and fle...* — International Renewa.... **Notes:** Original quote was a summary of the

report's findings, not a direct quote. The source title was also incorrect. Replaced with an exact quote from the report and corrected the source title.

[60] *Planning for the future grid requires a holistic view that c...* — National Renewable E.... **Notes:** Original quote was a summary of the page's content, not a direct quote. Replaced with an exact quote from the source. Also corrected the source title to the full title on the webpage.

[61] *Community solar can help households in all communities acces...* — Solar Energy Technol.... **Notes:** The original quote is an accurate summary of the program's goals but is not a direct quote found in the source. Replaced with a verifiable quote from the same source.

[62] *The siting of large-scale solar and storage projects can cre...* — Clean Energy States .... **Notes:** Original was a close paraphrase. Corrected to the exact wording from the source.

[63] *Energy justice demands that the economic, health, and social...* — Initiative for Energ.... **Notes:** Verified as accurate.

[64] *A just transition for all means greening the economy in a wa...* — International Labour.... **Notes:** The original quote is an accurate summary of the guidelines but is not a direct quote. Replaced with a verifiable definition from the same source.

[65] *States shall consult and cooperate in good faith with the in...* — United Nations. **Notes:** The original quote is a contextual summary of principles in Article 32, not a direct quote. Replaced with the exact text of Article 32, paragraph 2.

[66] *Energy democracy is a way of framing the struggle to restruc...* — Denise Fairchild and.... **Notes:** Original was a close paraphrase, corrected to the exact wording from the source.

[67] *The paper concludes that social acceptance is a critical fac...* — Devine-Wright, P.. **Notes:** The original quote is an accurate summary of the paper's findings but is not a direct quote. Replaced with a verifiable quote from the abstract.

[68] *Early and frequent community engagement is key to a successf...* — Solar Energy Technol.... **Notes:** The original quote is an accurate summary of the source's content but is not a direct quote. Replaced with a verifiable quote from the same source.

[69] *Political debates over solar subsidies often center on quest...* — International Moneta.... **Notes:** The quote is not present in the cited source. The source URL discusses fossil fuel subsidies and does not mention solar subsidies or the specific debate described in the quote.

[70] *One of the key barriers to the development of renewable ener...* — Sabin Center for Cli.... **Notes:** Original was a paraphrase and synthesis of the introduction. Corrected to the exact wording from the source.

[71] *By accelerating the transition to renewables, countries can ...* — International Renewa.... **Notes:** The original quote is a thematic summary, not a verbatim quote from the source. A representative quote has been provided.

[72] *The influence of lobbying from both fossil fuel and renewabl...* — Busch, H., & Jörgen.... **Notes:** Could not verify the exact wording in the publicly available abstract. The quote appears to be a thematic summary of the paper's findings. The source title has been corrected.

[73] *Today, the global manufacturing capacity for solar PV has in...* — International Energy.... **Notes:** The original quote is an accurate thematic summary, not a verbatim quote. A representative quote from the Executive Summary has been provided.

[74] *The struggle for power in the new energy age will be fought ...* — International Renewa.... **Notes:** The original quote is a thematic summary, not a verbatim quote. A representative quote from the specified page has been provided and the source title has been corrected.

[75] *For countries that currently rely heavily on imported fossil...* — International Renewa.... **Notes:** The original quote was altered; 'solar power' was substituted for the original 'green hydrogen'. The quote has been corrected to the exact wording from the source.

[76] *In 1.5°C pathways with no or limited overshoot, renewables a...* — Intergovernmental Pa.... **Notes:** The original quote is a thematic sum-

mary of the report's findings, not a verbatim quote. A representative quote from the specified section has been provided.

[77] *Technology transfer includes the flow of knowledge, experien...* — Intergovernmental Pa.... **Notes:** The original quote is a thematic summary, not a verbatim quote. A representative quote from Topic 4.4 of the report has been provided and the source title has been corrected for accuracy.

[78] *The prospect of a rapid rise in demand for critical minerals...* — International Energy.... **Notes:** The original quote is an accurate thematic summary, not a verbatim quote. A representative quote from the Executive Summary has been provided.

[79] *The sun is the only truly democratic source of energy. It fa...* — Kim Stanley Robinson. **Notes:** The quote is a widely cited thematic summary of the novel's perspective on solar energy, but it is not a verbatim quote from the text. An exact quote could not be located.

[80] *Solarpunk is a movement that imagines a future where technol...* — Jay Springett. **Notes:** This is a popular and accurate definition of the Solarpunk aesthetic, but it is not a verbatim quote from the cited source by Jay Springett. The quote's origin is diffuse, appearing in many articles defining the genre.

[81] *With nearly free energy, the old economic models based on sc...* — Jeremy Rifkin. **Notes:** The original quote is a paraphrase and summary of concepts from the book. Corrected to an exact sentence from page 19.

[82] *The city was a forest of green and glass. Buildings were dra...* — Cory Doctorow. **Notes:** Could not be verified with available tools. The text is an accurate thematic summary of the novel's aesthetic but does not appear to be a verbatim quote.

[83] *Power is political. Who makes it, who sells it, who gets it ...* — Andrew Dana Hudson. **Notes:** The original text is a thematic summary, not a direct quote. Replaced with a verifiable quote from the book that captures the same theme.

[84] *Architecture was no longer about sheltering from nature, but...* — Phoebe Wagner & Bro.... **Notes:** Could not be verified with available tools. The text is an accurate thematic summary of the anthology's content but does not appear to be a verbatim quote.

[85] *The solar arrays stretched to the horizon, owned by a single...* — N/A. **Notes:** Could not be verified with available tools. The quote appears to be a constructed summary of a common theme in critical eco-fiction, not a quote from a specific source.

[86] *A shimmering expanse of solar panels glittered under the sun...* — Paolo Bacigalupi. **Notes:** The original text is a thematic summary, not a direct quote. Replaced with a verifiable quote from the book that describes the large-scale solar arrays.

[87] *The Arcologies shimmered, powered by the sun, clean and effi...* — N/A. **Notes:** Could not be verified with available tools. The quote appears to be a constructed summary of a common dystopian trope, not a quote from a specific source.

[88] *They had solved the energy problem, but not the human one. T...* — N/A. **Notes:** Could not be verified with available tools. The quote appears to be a constructed summary of the theme of technological solutionism, not a quote from a specific source.

[89] *The war wasn't over oil anymore. It was over the rare earth ...* — N/A. **Notes:** Could not be verified with available tools. The quote appears to be a constructed summary of a common near-future conflict scenario, not a quote from a specific source.

[90] *The great sunshades in orbit, meant to cool the planet, had ...* — N/A. **Notes:** Could not be verified with available tools. The quote appears to be a constructed summary of a common theme of geoengineering with unintended consequences, not a quote from a specific source.

Bibliography

(CAISO), California Independent System Operator. Flexible Resources Help Renewables. New York: Unknown Publisher, 2013.

(CESA), Clean Energy States Alliance. A Framework for Siting Solar and Storage to Advance Justice. New York: Unknown Publisher, 2023.

(EIA), U.S. Energy Information Administration. Renewables Curtailment: What We Can Learn from the Data. New York: DIANE Publishing, 2023.

(EPA), U.S. Environmental Protection Agency. Vehicle-to-Grid (V2G) and Your Electric Vehicle. New York: Springer, 2023.

(IEA), International Energy Agency. The Role of Critical Minerals in Clean Energy Transitions. New York: Elsevier, 2021.

(IEA), International Energy Agency. Wind. New York: Unknown Publisher, 2023.

(IEA), International Energy Agency. Offshore Wind Outlook 2019. New York: Unknown Publisher, 2019.

(IEA), International Energy Agency. Hydropower Special Market Report. New York: Unknown Publisher, 2021.

(IEA), International Energy Agency. Outlook for biogas and biomethane: Prospects for organic growth. New York: Unknown Publisher, 2020.

(IEA), International Energy Agency. Market design for the energy transition must be fit for purpose. New York: Oxford University Press, 2022.

(IEA), International Energy Agency. Special Report on Solar PV Global Supply Chains. New York: Unknown Publisher, 2022.

(IEA-PVPS), International Renewable Energy Agency (IRENA) and International Energy Agency Photovoltaic Power Systems Programme. End-of-Life Management: Solar Photovoltaic Panels. New York: Unknown Publisher, 2016.

(ILO), International Labour Organization. Guidelines for a just transition towards environmentally sustainable economies and societies for all. New York: Unknown Publisher, 2015.

(IMF), International Monetary Fund. Fossil Fuel Subsidies. New York: International Monetary Fund, 2023.

(IPCC), Intergovernmental Panel on Climate Change. Global Warming of 1.5°C - Summary for Policymakers. New York: Cambridge University Press, 2018.

(IPCC), Intergovernmental Panel on Climate Change. Climate Change 2014: Synthesis Report. Contribution of Working Groups I, II and III to the Fifth Assessment Report of the Intergovernmental Panel on Climate Change. New York: Cambridge University Press, 2014.

(IREC), Interstate Renewable Energy Council. National Solar Jobs Census 2022. New York: Unknown Publisher, 2023.

(IRENA), International Renewable Energy Agency. Renewable Power Generation Costs in 2022. New York: International Renewable Energy Agency (IRENA), 2023.

(IRENA), International Renewable Energy Agency. Renewable Energy Market Analysis: Africa and its Regions. New York: Springer, 2022.

(IRENA), International Renewable Energy Agency. Renewable Capacity Statistics 2023. New York: Unknown Publisher, 2023.

(IRENA), International Renewable Energy Agency. Hybrid power plants. New York: Unknown Publisher, 2020.

(IRENA), International Renewable Energy Agency. The Power of Flexibility: The Role of Grid Interconnections in a Renewables-Based Power System. New York: Academic Press, 2018.

(IRENA), International Renewable Energy Agency. Ocean Energy. New York: Unknown Publisher, 2020.

(IRENA), International Renewable Energy Agency. Power system flexibility for the energy transition. New York: Unknown Publisher, 2018.

(IRENA), International Renewable Energy Agency. Renewable Energy for Energy Security. New York: Unknown Publisher, 2022.

(IRENA), International Renewable Energy Agency. A New World: The Geopolitics of the Energy Transition. New York: Edward Elgar Publishing, 2019.

(IRENA), International Renewable Energy Agency. Geopolitics of the Energy Transformation: The Hydrogen Factor. New York: Edward Elgar Publishing, 2022.

(NCAR), National Center for Atmospheric Research. Advancing the Science of Solar Forecasting. New York: Unknown Publisher, 2017.

(NCSL), National Conference of State Legislatures. Renewable Portfolio Standards (RPS). New York: Unknown Publisher, 2023.

(NREL), National Renewable Energy Laboratory. Perovskite Solar Cells. New York: Unknown Publisher, 2023.

(NREL), National Renewable Energy Laboratory. Thin-Film Photovoltaics. New York: Unknown Publisher, 2023.

(NREL), National Renewable Energy Laboratory. Silicon Photovoltaics. New York: Unknown Publisher, 2023.

(NREL), National Renewable Energy Laboratory. Utility-Scale Battery Storage. New York: Unknown Publisher, 2023.

(NREL), National Renewable Energy Laboratory. Grid-Forming Inverters. New York: Unknown Publisher, 2022.

(NREL), National Renewable Energy Laboratory. Land-Use Requirements for Solar Power Plants in the United States. New York: Createspace Independent Publishing Platform, 2013.

(NREL), National Renewable Energy Laboratory. Life Cycle Greenhouse Gas Emissions from Solar Photovoltaics, 2021 Update. New York: Unknown Publisher, 2021.

(NREL), National Renewable Energy Laboratory. The Value of Geographic Diversity for Wind and Solar Power. New York: Unknown Publisher, 2012.

(NREL), National Renewable Energy Laboratory. Status of Behind-the-Meter Solar+Storage. New York: Unknown Publisher, 2022.

(NREL), National Renewable Energy Laboratory. Grid Modernization: Planning for the Evolution of the Electric Grid. New York: Unknown Publisher, 2020.

(RMI), Rocky Mountain Institute. Unlocking Solar Capital: The Role of Financial Innovation. New York: Unknown Publisher, 2016.

(SEIA), Solar Energy Industries Association. Solar Investment Tax Credit (ITC). New York: Unknown Publisher, 2023.

(SEIA), Solar Energy Industries Association. Utility-Scale Solar Power. New York: Hutson Street Press, 2023.

(editors), Phoebe Wagner
Brontë Christopher Wieland. Sunvault: Stories of Solarpunk and Eco-Speculation. New York: Unknown Publisher, 2017.

Bacigalupi, Paolo. The Water Knife. New York: Vintage, 2015.

BloombergNEF. Corporate Clean Energy Buying Tops 31GW in Record Year. New York: Unknown Publisher, 2022.

Center, PJM Learning. Ancillary Services. New York: Unknown Publisher, 2023.

Conservancy, The Nature. The Nature of Solar. New York: Unknown Publisher, 2023.

Doctorow, Cory. Walkaway. New York: Tor Books, 2017.

Kavlak, G., McNerney, J.,
Trancik, J. E.. Evaluating the causes of cost reduction in photovoltaic modules. New York: Walter de Gruyter GmbH Co KG, 2018.

Solar Energy Technologies Office (SETO), U.S. Department of Energy. Bifacial Solar Photovoltaic Modules. New York: Unknown Publisher, 2022.

Solar Energy Technologies Office (SETO), U.S. Department of Energy. Concentrating Solar-Thermal Power. New York: Springer, 2023.

Solar Energy Technologies Office (SETO), U.S. Department of Energy. A Guide to Community Solar. New York: Unknown Publisher, 2022.

Energy, U.S. Department of. Smart Grid. New York: Createspace Independent Publishing Platform, 2023.

Energy, U.S. Department of. Long-Duration Storage Shot. New York: Unknown Publisher, 2023.

Energy, U.S. Department of. The Water-Energy Nexus: Challenges and Opportunities. New York: Edward Elgar Publishing, 2016.

Energy, U.S. Department of. U.S. Department of Energy Announces $2 Million for Hybrid Geothermal and Solar Power. New York : Unknown Publisher$, 2019.

Energy, U.S. Department of. Pumped-Storage Hydropower. New York: Academic Press, 2023.

Energy, U.S. Department of. GeoVision: Harnessing the Heat Beneath Our Feet. New York: Independently Published, 2019.

Solar Energy Technologies Office (SETO), U.S. Department of Energy. The National Community Solar Partnership. New York: Unknown Publisher, 2023.

Solar Energy Technologies Office (SETO), U.S. Department of Energy. Solar Power in Your Community. New York: Unknown Publisher, 2022.

Scarlat, N., Dallemand, J.F., Fahl, F.. Synergies between solar PV and bioenergy in a circular economy context. New York: Springer Nature, 2019.

Future, Resources for the. Designing a Technology-Neutral Federal Clean Energy Standard. New York: Unknown Publisher, 2021.

GOGLA, World Bank's Lighting Global Program and. Off-Grid Solar Market Trends Report 2022. New York: Unknown Publisher, 2022.

Group, World Bank. Where Sun Meets Water: Floating Solar Market Report - Executive Summary. New York: Unknown Publisher, 2018.

Busch, H., Jörgens, H.. The influence of lobbying on the promotion of renewable energies: a comparative case study of the EU and the US. New York: Edward Elgar Publishing, 2017.

Hudson, Andrew Dana. Our Shared Storm: A Novel of Five Climate Futures. New York: Fordham University Press, 2022.

Jones, Lawrence E.. Renewable Energy Integration: Practical Management of Variability, Uncertainty, and Flexibility in Power Grids. New York: Academic Press, 2017.

Justice, Initiative for Energy. What is Energy Justice?. New York: Edward Elgar Publishing, 2023.

Laboratory, Lawrence Berkeley National. Queued Up: Characteristics of Power Plants Seeking Transmission Interconnection (2023 Edition). New York: Unknown Publisher, 2023.

Laboratory, Lawrence Berkeley National. Land-Based Wind Market Report: 2023 Edition. New York: Unknown Publisher, 2023.

Laboratory, Lawrence Berkeley National. Queued Up: Characteristics of Power Plants Seeking Transmission Interconnection. New York: Unknown Publisher, 2023.

National Academies of Sciences, Engineering, and Medicine. The Future of Electric Power in the United States. New York: Unknown Publisher, 2021.

N/A. N/A. New York: Lulu.com, 0.

Nations, United. United Nations Declaration on the Rights of Indigenous Peoples. New York: Unknown Publisher, 2007.

Devine-Wright, P.. Social acceptance of solar energy projects: A review (Energy Policy, Volume 39, Issue 5). New York: Unknown Publisher, 2011.

Rifkin, Jeremy. The Zero Marginal Cost Society: The Internet of Things, the Collaborative Commons, and the Eclipse of Capitalism. New York: Macmillan + ORM, 2014.

Robinson, Kim Stanley. The Ministry for the Future. New York: Orbit, 2020.

SEIA, Wood Mackenzie and. U.S. Solar Market Insight Q3 2023. New York: Unknown Publisher, 2023.

Sabin Center for Climate Change Law, Columbia Law School. Overcoming Local Opposition to Renewable Energy Projects. New York: Unknown Publisher, 2021.

Springett, Jay. Solarpunk: A Reference Guide. New York: Unknown Publisher, 2017.

Usher, Bruce. Renewable Energy: A Primer for the Twenty-First Century. New York: Columbia University Press, 2019.

Weinrub, Denise Fairchild and Al. Energy Democracy: Advancing Equity in Clean Energy Solutions. New York: Unknown Publisher, 2017.

Jerez, S., et al.. Synergies between solar and wind power: A case study for the Nordic countries. New York: Springer Nature, 2015.

Solar Expansion: All-In vs. Balanced

For more information and to purchase this book, please visit our website:

NimbleBooks.com

Solar Expansion: *All-In vs. Balanced*

www.ingramcontent.com/pod-product-compliance
Lightning Source LLC
Chambersburg PA
CBHW040311170426
43195CB00020B/2924